Mathematics
WARM-UPS
Grade 7

WALCH EDUCATION

1 2 3 4 5 6 7 8 9 10
ISBN 978-0-8251-7148-2

J. Weston Walch, Publisher
Portland, ME 04103
www.walch.com
Printed in the United States of America

Table of Contents

Introduction

Mathematics Warm-Ups for Common Core State Standards, Grade 7 is organized into five sections, composed of the domains for Grade 7 as designated by the Common Core State Standards Initiative. Each warm-up addresses at least one of the standards within the following domains:

- Ratios and Proportional Relationships
- The Number System
- Expressions and Equations
- Geometry
- Statistics and Probability

The Common Core Mathematical Practices standards are another focus of the warm-ups. All of the problems require students to "make sense of problems and persevere in solving them," "reason abstractly and quantitatively," and "attend to precision." Students must "look for and make use of structure" when finding inputs and outputs in a function machine. Students have opportunities to "use appropriate tools strategically" when they use 10×10 grids to examine fractions and decimals or coin tosses to explore sample area and probability. A full description of these standards can be found at http://www.walch.com/CCSS/00006.

The warm-ups are organized by domains rather than by level of difficulty. Use your judgment to select appropriate problems for your curriculum.* The problems are not necessarily meant to be completed in consecutive order—some are stand-alone, some can launch a topic, some can be used as journal prompts, and some refresh students' skills and concepts. All are meant to enhance and complement your Grade 7 mathematics program. They do so by providing resources for those short, 5- to 15-minute interims when class time might otherwise go unused.

* You may select warm-ups based on particular standards using the Standards Correlations table.

About the CD-ROM

Mathematics Warm-Ups for Common Core State Standards, Grade 7 is provided in two convenient formats: an easy-to-use reproducible book and a ready-to-print PDF on a companion CD-ROM. You can photocopy or print activities as needed, or project them on a screen via your computer.

The depth and breadth of the collection give you the opportunity to choose specific skills and concepts that correspond to your curriculum and instruction. Use the table of contents and the standards correlations to help you select appropriate tasks.

Suggestions for use:

- Choose an activity to project or print out and assign.
- Select a series of activities. Print the selection to create practice packets for learners who need help with specific skills or concepts.

Standards Correlations

Mathematics Warm-Ups for Common Core State Standards, Grade 7 is correlated to five domains of CCSS Grade 7 mathematics. The page numbers, titles, and standard numbers are included in the table that follows. The full text of the CCSS mathematics standards for Grade 7 can be found in the Common Core State Standards PDF at http://www.walch.com/CCSS/00001.

Page number	Title	CCSS addressed
Ratios and Proportional Relationships		
1	Balancing a Milk Bottle	7.RP.1
2	Check It Out	7.RP.2a
3	Comparing Rectangles	7.RP.2b
4	Input and Output	7.RP.2b
5	Pizza Pizza Pizza	7.RP.2b
6	Tree Height	7.RP.2b
7	Vanishing Wetlands	7.RP.3
8	Percent Increase or Decrease	7.RP.3
9	Bread Prices	7.RP.3
The Number System		
10	Chip-Board Integers I	7.NS.1c
11	Working with Integers I	7.NS.1d
12	Working with Integers II	7.NS.1d
13	Chip-Board Integers II	7.NS.1d
14	Chip-Board Integers III	7.NS.1d
15	Integer Practice	7.NS.1d
16	Make a True Sentence	7.NS.1d, 7.NS.2c
17	10×10 Grids	7.NS.2d

(*continued*)

Page number	Title	CCSS addressed
18	It's Completely Rational	7.NS.3
19	Four 4s	7.NS.3
20	Numbering Pages	7.NS.3
21	A Circle of Students	7.NS.3
22	Compressing Trash	7.NS.3
Expressions and Equations		
23	Perimeter Problem	7.EE.1
24	Thinking Around the Box	7.EE.1
25	Simplifying Expressions	7.EE.1
26	Guess and Check	7.EE.4
Geometry		
27	Scale It Up	7.G.1
28	Hiking a National Park	7.G.1
29	Can You Do It?	7.G.2
30	Angles and Triangles	7.G.2
31	Slice It Up	7.G.3
32	Considering Pizza Costs	7.G.4
33	Pondering Pizza Prices	7.G.4
34	Circles and Squares	7.G.4
35	Perimeter Expressions	7.G.4
36	Cynthia's Cylinder	7.G.4
37	Complementary and Supplementary Angles	7.G.5
38	Volume and Surface Area I	7.G.6
39	Volume and Surface Area II	7.G.6
40	A Rectangular Box	7.G.6
41	Area and Perimeter	7.G.6
42	Storing DVDs	7.G.6

(*continued*)

Page number	Title	CCSS addressed
43	Storing Trash	7.G.6
44	Beach Volleyball	7.G.6
	Statistics and Probability	
45	Pasta Primavera or Chicken Teriyaki?	7.SP.1
46	In the Median	7.SP.4
47	Testing, Testing	7.SP.4
48	Ring Toss	7.SP.5
49	Coin Chances	7.SP.6
50	Pairing Socks	7.SP.8a
51	Tossing Coins	7.SP.8a, 7.SP.8b
52	Ice Cream Cones	7.SP.8a, 7.SP.8b
53	Spinning for Numbers	7.SP.8b

Balancing a Milk Bottle

An American named Ashrita Furman holds more Guinness World Records than any other person. In April 1998, he walked 81 miles in 23 hours, 35 minutes while balancing a milk bottle on his head. How fast did he walk in miles per hour?

Check It Out

The graph below illustrates the distance a car traveled at 25 miles per hour. Examine the graph closely and describe how it looks to a partner.

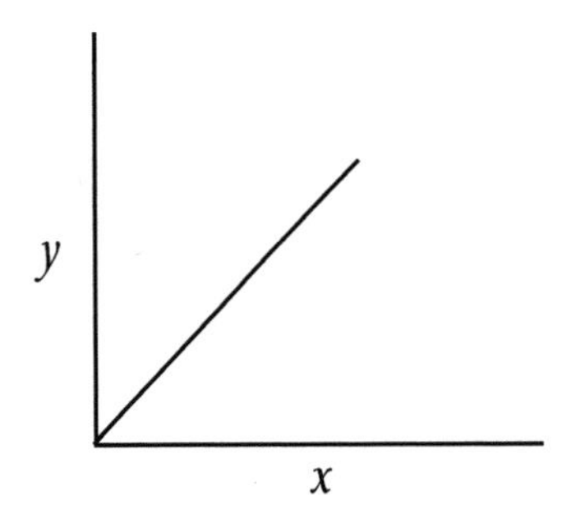

What information lies on the x-axis?

What information lies on the y-axis?

Does it make sense that the graph looks as it does? Explain. Write your observations below.

Comparing Rectangles

Examine the two rectangles below. Notice that the larger rectangle's length and width are three times the smaller rectangle's length and width.

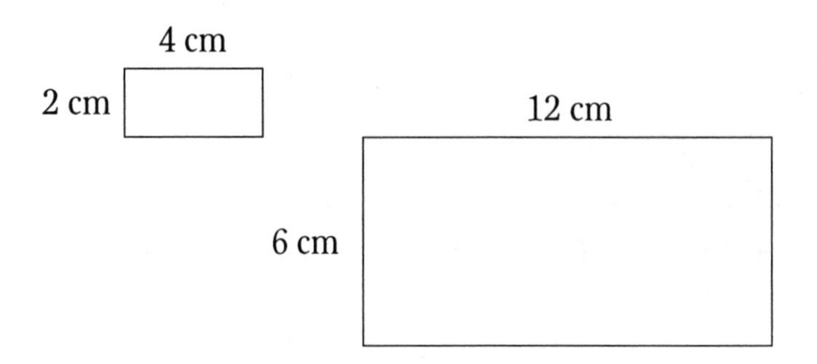

Answer the following questions.

1. What is the area of the smaller rectangle?

2. What is the area of the larger rectangle?

3. How many times larger is the area of the larger rectangle than the smaller rectangle? Is this what you expected? Explain.

Input and Output

Consider the tables below. Figure out the rule that takes the input value and gives the corresponding output value. Describe each rule in words.

Example:

Input	1	2	3	4	5
Output	4	5	6	7	8

Rule: Add 3 to the input to get the output.

1.

Input	5	7	9	11	13
Output	10	14	18	22	26

2.

Input	10	20	30	40	50
Output	6	16	26	36	46

3.

Input	24	28	32	36	40
Output	12	14	16	18	20

4.

Input	0	2	4	6	8
Output	3	7	11	14	19

Pizza Pizza Pizza

At Suki's Pizza Parlor, each slice of pizza costs $1.50. Aaron had been studying adding decimal numbers in math class, and so while waiting for his pizza, he made the chart below.

Number of slices	Price
1	$1.50
2	$3.00
3	$4.50
4	$6.00
5	$7.50

Use Aaron's chart to answer the following questions.

1. How would you describe how the numbers in the first column change?

2. How would you describe how the numbers in the second column change?

Tree Height

Use what you know about proportional relationships to answer the questions that follow. Show your work for each problem.

1. Justine, who is 60 inches tall, casts an 8-foot shadow. At the same time and place, how long a shadow would a 35-foot tree cast?

2. If another tree nearby casts a 40-foot shadow, how tall is this tree?

Vanishing Wetlands

Bonita works for the Desoto Park Service. The Little Otter Wetland Area that she monitors has had drought conditions recently. She is preparing a report on the drought for the park service. The grid below represents a model for the change in area of the wetland. What was the percent change in wetland area from August 2010 to August 2011? Explain your thinking.

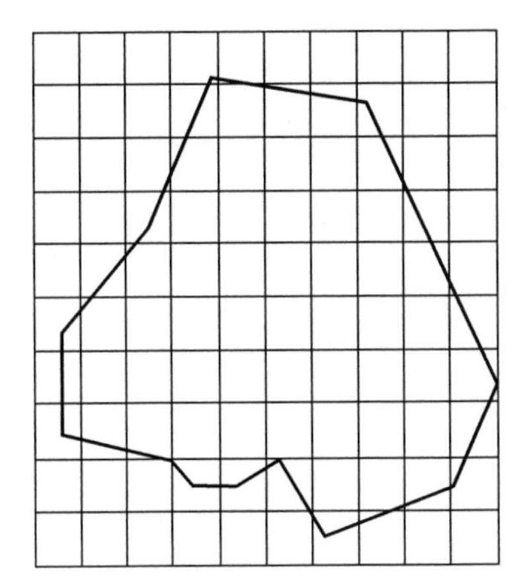

August 2010 model

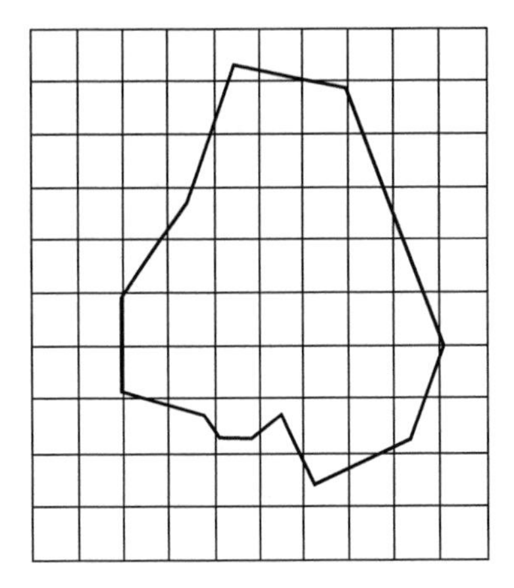

August 2011 model

Percent Increase or Decrease

Look at the sequences below. For each, tell whether there is growth or decay, identify the common ratio, and give the percent increase or decrease.

1. 43, 129, 387, 1,161, . . .

2. 90, 99, 108.9, 119.79, . . .

3. 1,800, 1,080, 648, 388.8, . . .

4. 17.8, 3.56, 0.712, 0.1424, . . .

5. 375, 142.5, 54.15, 20.577, . . .

Bread Prices

The price of bread changes from year to year due to factors such as the condition of the wheat harvest, the weather, and the inflation rate. Read each question and answer it based on the information provided.

1. According to the U.S. Bureau of Labor Statistics, a loaf of white bread cost $0.79 in 1995 and $1.04 in 2005. What is the percent increase in price for this 10-year period?

2. If the trend continues, what would be the predicted price for a loaf of white bread in 2015?

3. If the price in 2010 was $1.37, is the trend continuing at the same percentage rate? Explain your thinking.

NAME:

THE NUMBER SYSTEM
CCSS 7.NS.1c

Chip-Board Integers I

Explain how you might show the operation (–6) – (–12) = ? on a chip board or a series of chip boards such as the ones pictured below. Draw black circles, or chips, to represent negative values. Draw white circles, or chips, to represent positive values. A black chip and a white chip together represent zero.

Chip board 1

Chip board 2

Chip board 3

Working with Integers I

Using appropriate arrow notations on number lines such as the one pictured below, represent the expressions given and find the results. Use a new number line for each sentence.

1. $-8 + 12 = ?$

2. $(-7) + (-3) = ?$

3. $5 + (-9) = ?$

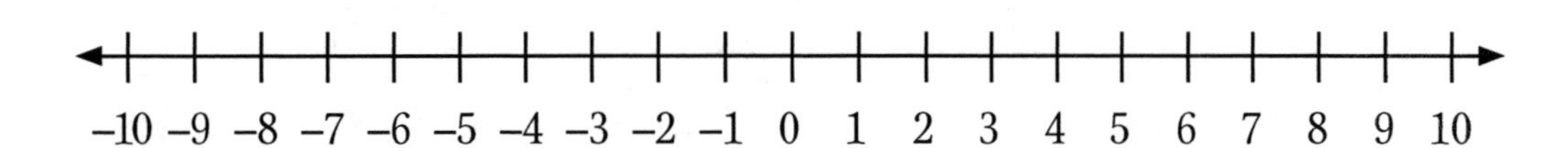

Working with Integers II

Look at each diagram below. Write a true sentence showing the integer calculations pictured.

1.

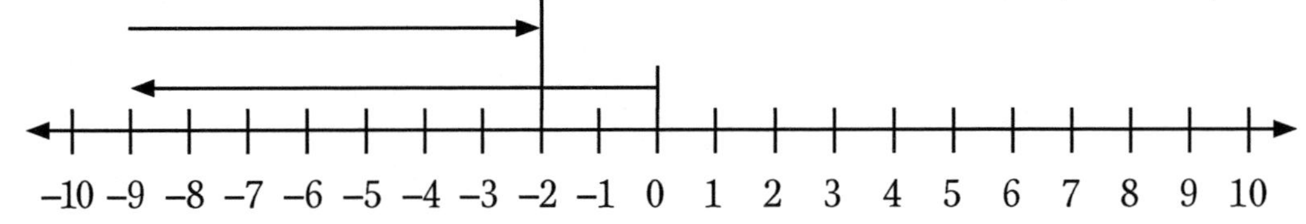

2.

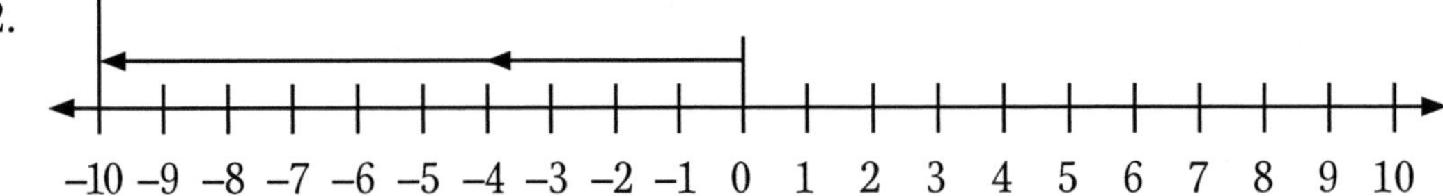

3.

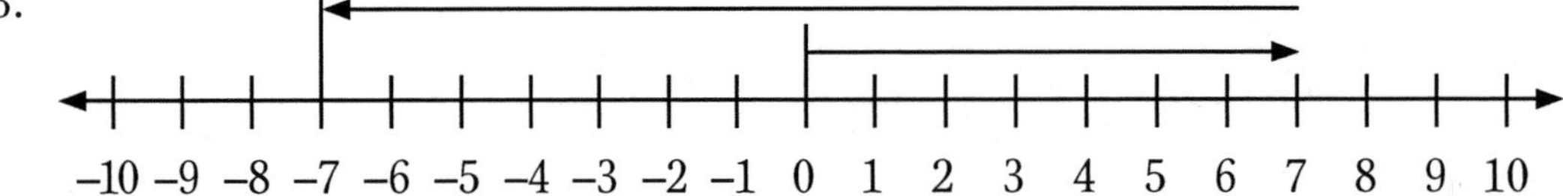

Chip-Board Integers II

Andrew and Brittany are exploring integers by drawing representations using black and white circular chips. The white chips represent positive numbers. The black chips represent negative numbers. Write a number sentence to symbolize each set of chip boards that they have drawn.

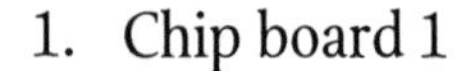

1. Chip board 1

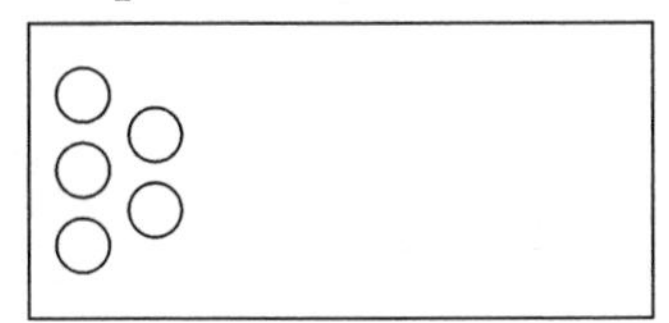

Chip board 2

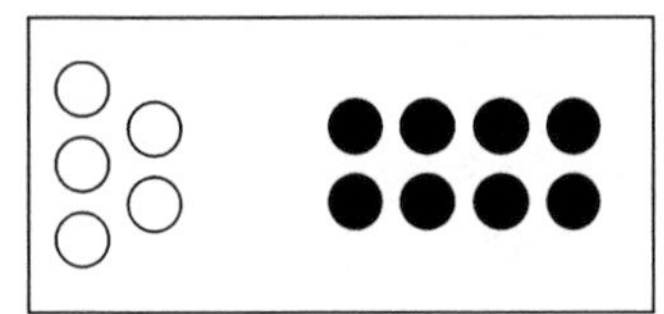

Chip board 3

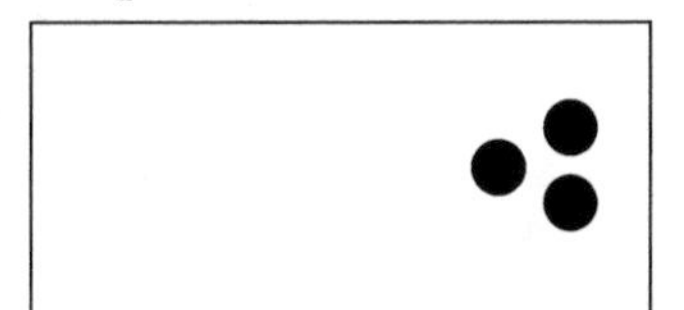

2. Chip board 1

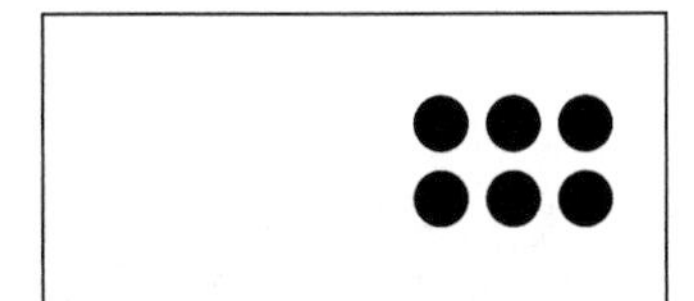

Chip board 2

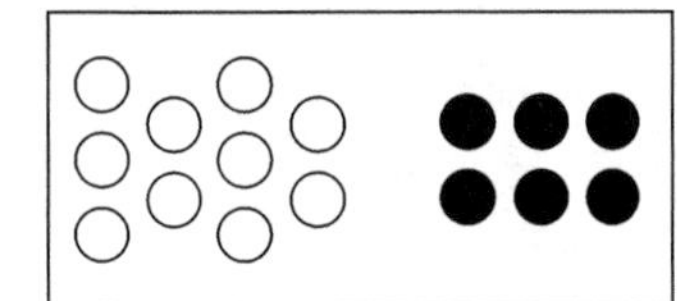

Chip board 3

3. Chip board 1

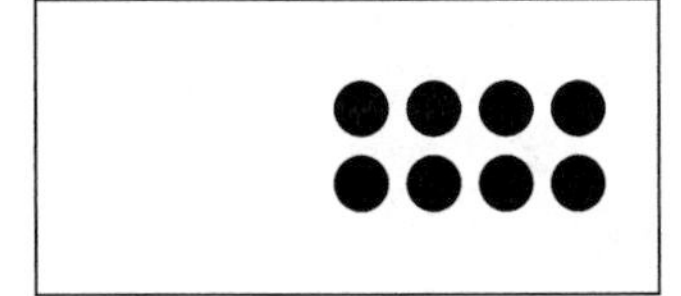

Chip board 2

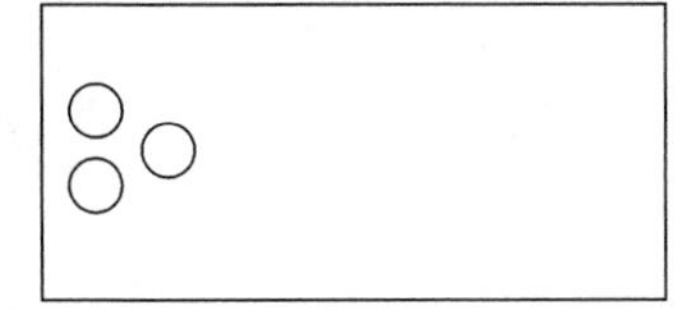

Chip board 3

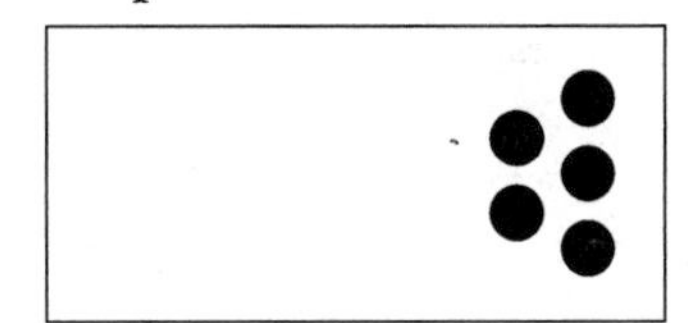

Chip-Board Integers III

Diagram and solve the following integer operations. Use chip boards such as the ones pictured below. Use black circles, or chips, to represent negative values. Use white circles, or chips, to represent positive values. Use a series of two or three chip boards for each sentence.

1. $(-5) + (9) = ?$

2. $12 + (-7) = ?$

3. $(-8) - (-5) = ?$

Chip board 1

Chip board 2

Chip board 3

Integer Practice

Find the missing value in each sentence below.

1. $(-6) + (-12) = ?$

2. $25 - (-8) = ?$

3. $? + 17 = (-4)$

4. $11 + ? = 6$

5. $\frac{3}{4} - ? = 1$

6. $(-7) + ? = 0$

7. $2\frac{1}{2} + ? = \frac{1}{2}$

8. $? - (-5) = 0$

9. $-3.5 - \frac{7}{2} = ?$

10. $(-5.6) - ? = (-3.4)$

11. $? + 7\frac{2}{5} = 3\frac{3}{5}$

12. $-\frac{87}{10} + ? = 10$

Make a True Sentence

For the sentence below, insert mathematical symbols of any kind to make the sentence true. It is possible to make 2- or 3-digit numbers by not inserting any symbols between the numbers (for instance, by putting 5 and 6 together to make 56).

$$1 \quad 2 \quad 3 \quad 4 \quad 5 \quad 6 \quad 7 \quad 8 \quad 9 = 100$$

NAME:

THE NUMBER SYSTEM
CCSS 7.NS.2d

10 × 10 Grids

Adriana has shaded a portion of each 10 × 10 grid pictured below. What fraction of each has been shaded? Express the fraction as a decimal. What percent of each grid has been shaded?

1. 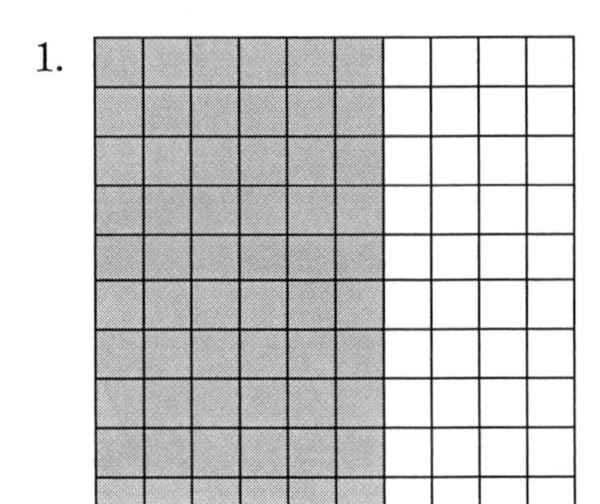

2. 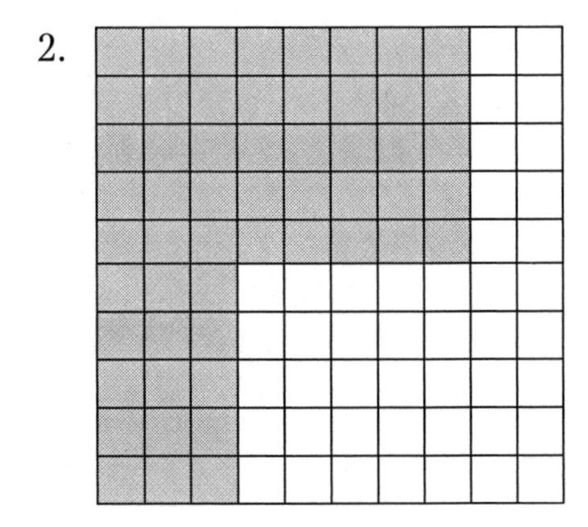

It's Completely Rational

Complete the following problems.

1. Ramon has 4.5 liters of soda. If he paid $2.02 per liter, how much did he pay for 4.5 liters?

2. Maggie's mom bought pizza for Maggie's birthday party. Each pizza was cut into 8 slices. After the party, there were $2\frac{3}{4}$ pizzas left. If Maggie's family ate half of the pizza that was left after the party, how much pizza remained? How many slices were left?

3. Julia bought a sandwich for $5.79 and a drink for $0.89. How much change would she get from a $10 bill?

Four 4s

Express each of the numbers 1 through 10 using only the following: the addition, subtraction, multiplication, and division symbols; four 4s; and parentheses, if necessary. For example, the number 12 can be expressed by 4 + 4 + 4 + 4. Show your work in the space below.

Numbering Pages

Read the scenario and use the information in it to answer the question that follows.

Dylan used 2,989 digits to number the pages of a book. How many pages are in the book? Remember that digits are the symbols 0, 1, 2, 3, 4, 5, 6, 7, 8, 9. Explain how you got your answer.

A Circle of Students

Every year on the first day of spring, all the students at Thoreau Middle School stand in a perfect, evenly spaced circle on the athletic field in honor of the vernal equinox. This year, Taylor notices that she is in seventh place on the circle. Directly opposite her in position number 791 is her friend Mattie. How many students make up the circle this year?

Compressing Trash

According to the U.S. Environmental Protection Agency's 2010 report, every person generates an average of 4.4 pounds of garbage per day. Of this, approximately 25% is recycled. If a cubic foot of compressed garbage weighs about 50 pounds, how much space would be needed for the garbage generated by a family of four in one year?

NAME:

EXPRESSIONS AND EQUATIONS
CCSS 7.EE.1

Perimeter Problem

Use the figure below to answer the questions that follow.

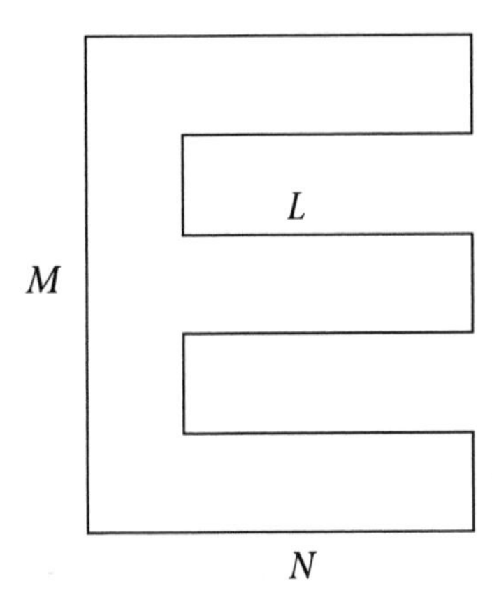

1. If $M = 5$ meters, $N = 4$ meters, and $L = 2.5$ meters, what is the perimeter of the figure?

2. Using the variables only, write two or more expressions that could represent the perimeter of the figure. Show that your expressions are equivalent.

3. Is there another way to represent the perimeter of the figure?

Thinking Around the Box

The rectangular box shown below has a front and back length of L meters. The side lengths are W meters. The height is H meters.

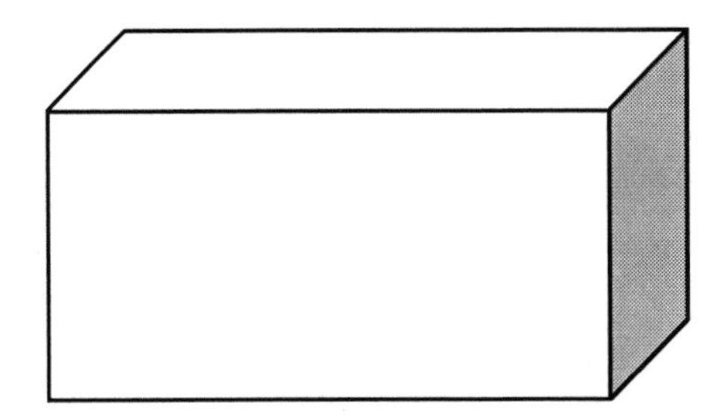

1. Write two equations that represent the sum, S, of all the edges of the rectangular box.

2. Write two equations that represent the surface area, A, of the rectangular box.

NAME:

EXPRESSIONS AND EQUATIONS
CCSS 7.EE.1

Simplifying Expressions

Write at least two more expressions that are equivalent to each given expression below.

1. $8(x - 5)$

2. $x(5x - 6) + 13x - 10$

3. $4(x + 5) - 3(2 - 4x)$

4. $2.5(12 - 2x) + 5(x + 1)$

5. $(3x^2 + 5x + 8) - (8x^2 + 2x - 5)$

Guess and Check

Some word problems can be solved using guess and check. Solve the following problems and be sure to check your answers to make sure they are correct. Write down all of your work in an organized way so someone else could follow your thinking, if needed.

1. Veronica is 2 years older than her brother Peter. The sum of their ages is 24. How old is Veronica, and how old is Peter?

2. Sofia has 4 times as many dollars as her friend José has. Together the two friends have $35. How much money do they each have?

Scale It Up

Read the scenario that follows. Use the information in it to draw a sketch and solve the problem.

Kurt had a square drawing that was 5 centimeters on each side. He wanted to enlarge the drawing so that it would be 4 times bigger. What would be the length of each side of Kurt's enlarged drawing? Sketch the enlarged drawing and label its side lengths.

NAME:

GEOMETRY
CCSS 7.G.1

Hiking a National Park

Use this map and a metric ruler to measure the distance from place to place. Help these hikers find their destination.

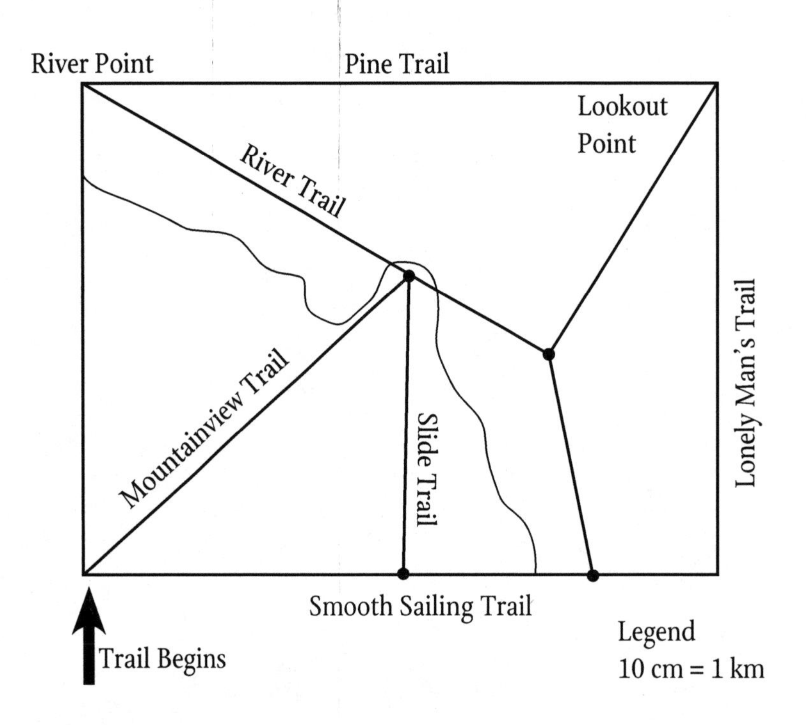

1. If the hikers walk down Mountainview Trail, turn left on River Trail, and stop at River Point, how many kilometers did they hike?

2. From River Point, they head along Pine Trail to Lookout Point. How far is it from River Point to Lookout Point?

3. If the hikers head down Lonely Man's Trail and turn right onto Smooth Sailing Trail to return to their starting point, how far will they have traveled from Lookout Point to their original starting point?

Can You Do It?

Observe the angle below. Now, get a compass and a ruler (NO protractor). Your challenge is to see if you can copy this angle only using a compass and a ruler. Can you find a way to do it? It *is* possible.

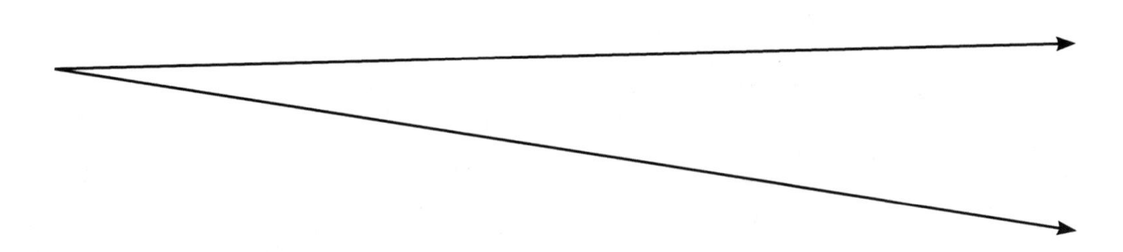

NAME:

GEOMETRY
CCSS 7.G.2

Angles and Triangles

In the diagram below, angle *EBF* is 50 degrees. It has been bisected (cut into two equal angles). The resulting angles are angle *EBG* and angle *FBG*. Segments *BE* and *BF* are the same length. Based on this information, answer the following questions.

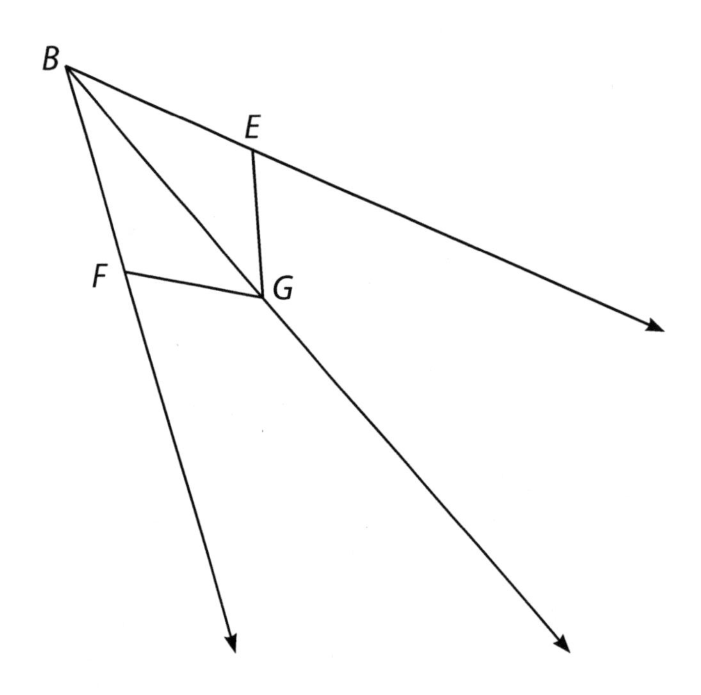

1. What is the measure of angle *EBG*? How do you know?

2. What is the measure of angle *FBG*? How do you know?

3. Are segments *EG* and *FG* the same length? Explain.

4. Are triangles *EGB* and *FGB* congruent? Explain.

NAME:

GEOMETRY
CCSS 7.G.3

Slice It Up

Look at the sketch of the cone below.

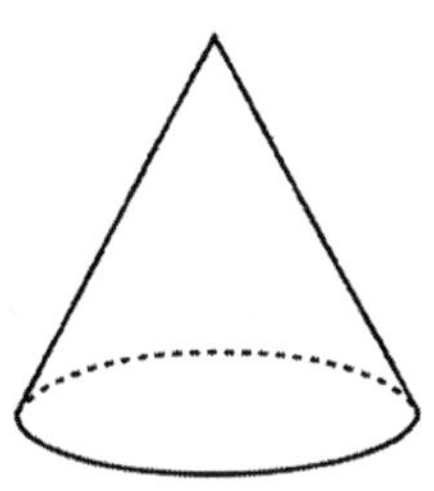

1. What shape is the bottom of the cone? Draw it in the space below.

2. If you took a horizontal slice of the cone about $\frac{1}{3}$ the way down, what shape would you get? Draw it in the space below.

3. If you took another horizontal slice of the cone about $\frac{2}{3}$ the way down, what shape you would get? Draw it in the space below.

4. If you took a vertical slice of the cone, straight through the center, what shape would you get? Draw it in the space below.

Considering Pizza Costs

Paisano's Pizza Parlor has just opened up across the street from Mei's apartment. Mei is thinking about inviting some friends over for pizza and ordering from Paisano's. She wants to get the most pizza for her money. The menu shows the following prices:

6-inch round pizza....................	$7.50
12-inch round pizza................	$12.00
18-inch round pizza................	$16.00

1. How many 6-inch pizzas have the same amount of pizza as one 12-inch pizza?

2. How many 6-inch pizzas have the same amount of pizza as one 18-inch pizza?

NAME:

GEOMETRY

CCSS 7.G.4

Pondering Pizza Prices

Antonio is the new owner of Paisano's Pizza Parlor. He wants to change the way his pizzas are priced. He is thinking about pricing them according to the diameter of each pizza. His cousin Vinnie recommends that he price them according to the size of the circumference of each pizza. His wife Bianca suggests pricing by the area of each pizza. Which method would you recommend to Antonio? Explain your reasoning.

NAME:

GEOMETRY
CCSS 7.G.4

Circles and Squares

Find the area of the shaded region for each figure below.

1. 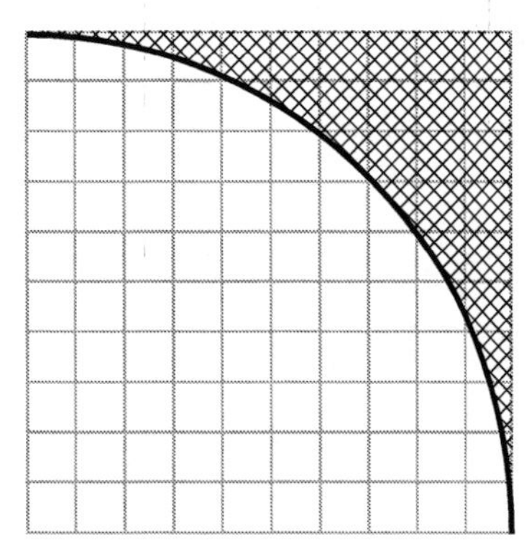

2. 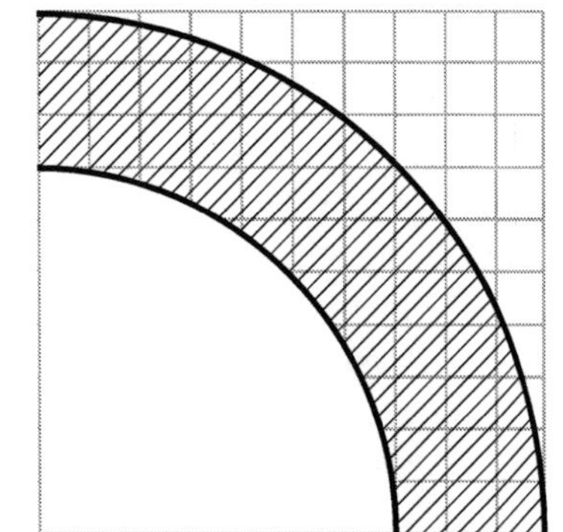

NAME:

GEOMETRY
CCSS 7.G.4

Perimeter Expressions

Rajah is working on her math homework. She has been given the sketch below, where r represents both the radius of the semicircle and the height of the rectangle that the semicircle is attached to. She is trying to decide on an expression that will represent the perimeter of the figure. She has written out the expressions below. Do they correctly represent the perimeter of the shape? Explain why or why not for each.

1. $r(2 + \pi)$

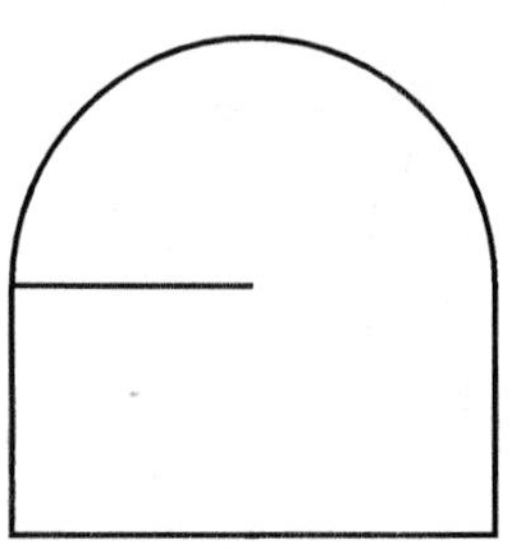

2. $4r + \pi r$

3. $r + 2r + r + 2\pi r$

4. $r(4 + \pi)$

5. $2r + \pi r$

NAME:

GEOMETRY
CCSS 7.G.4

Cynthia's Cylinder

Cynthia has drawn the model below to represent a cylindrical cardboard box that she needs to package a gift of lotion for her mother. If the package needs to be 8 inches tall and have a diameter of 3 inches, what is the area of the rectangle pictured? What is the total surface area of the box that she will create from her design?

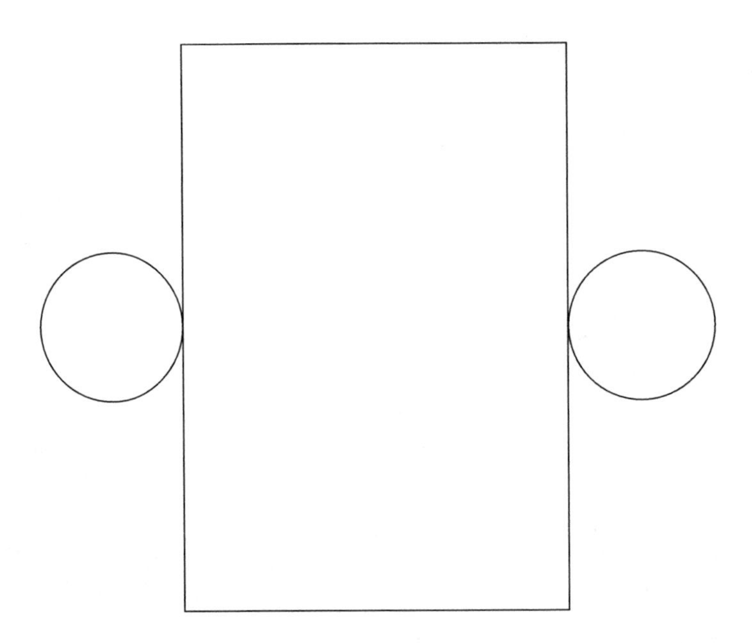

NAME:

GEOMETRY
CCSS 7.G.5

Complementary and Supplementary Angles

A right angle is a 90° angle. Two angles whose sum is 90° are complementary. Find and label the missing angle(s).

1.

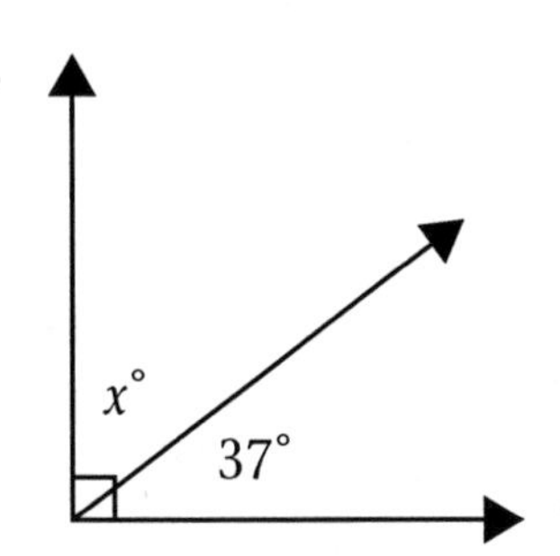

2.

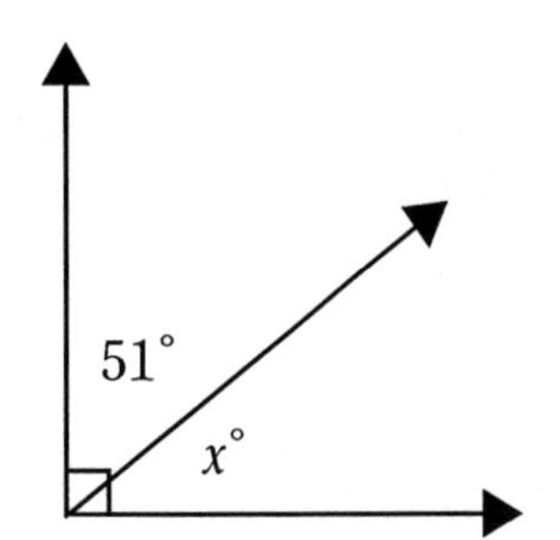

3.

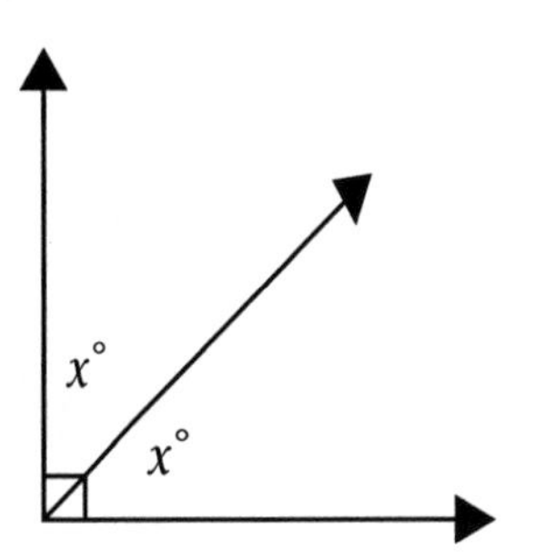

A straight angle is a straight line and measures 180°. Two angles whose sum is 180° are supplementary. Find and label the missing angle(s).

4.

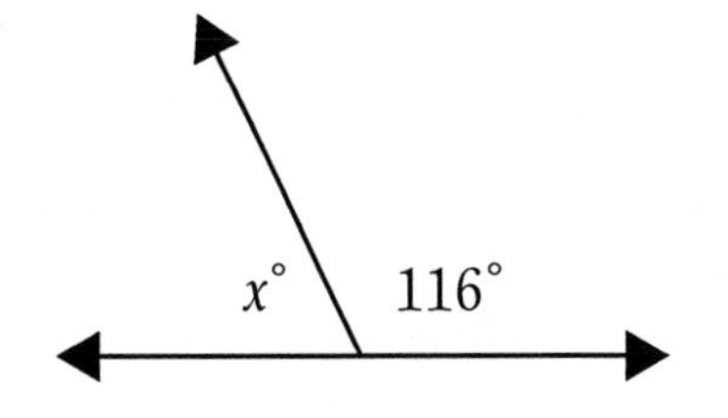

5.

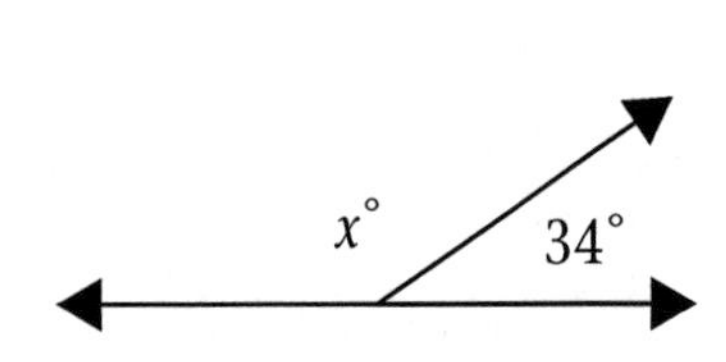

6.

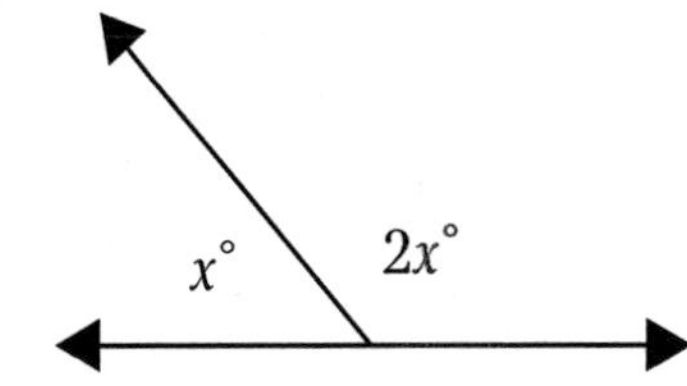

Volume and Surface Area I

The drawing below is a flat pattern or a net. When folded, it creates a box in the shape of a rectangular prism. Draw a sketch of the box and determine its total surface area and volume.

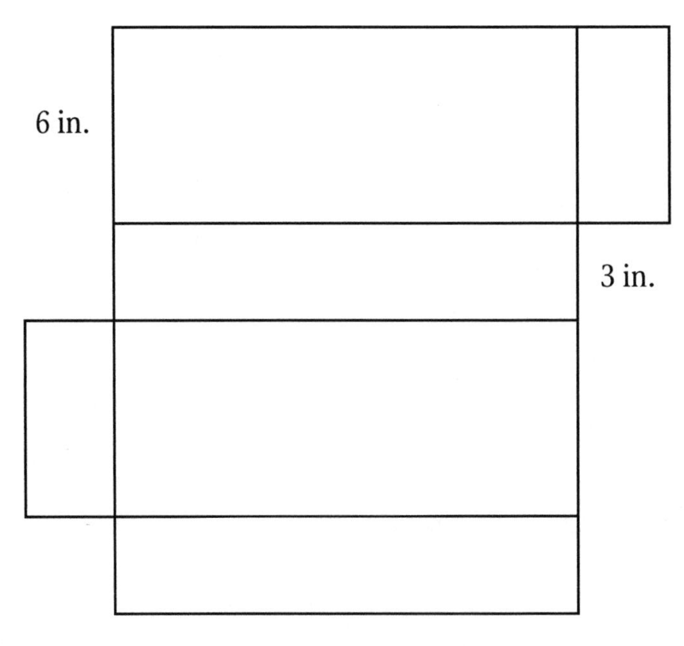

NAME:

GEOMETRY
CCSS 7.G.6

Volume and Surface Area II

The drawing below represents an open-topped rectangular prism with a length of 13 feet, a height of 6 feet, and a width of 5 feet. Draw a flat pattern or a net for the figure. Determine the surface area and volume of the box.

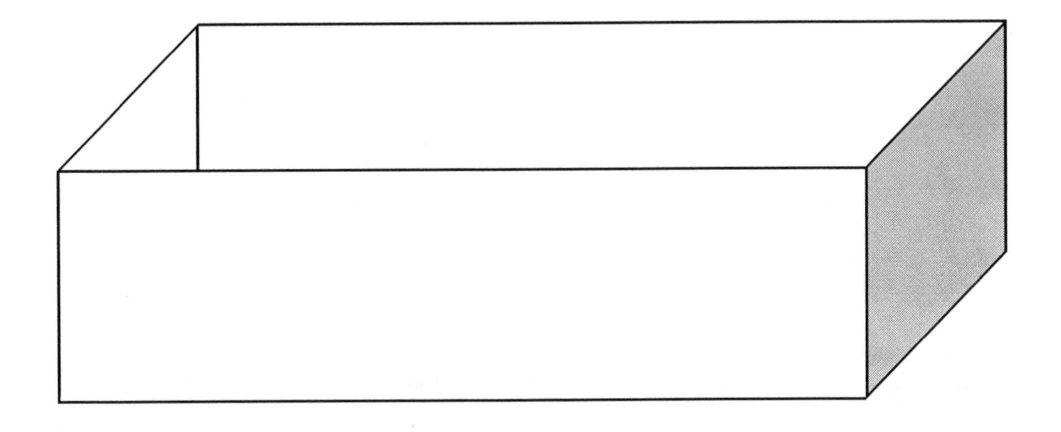

A Rectangular Box

In the space below, make a sketch of a rectangular box with a base of 3 inches by 5 inches and a height of 7 inches. Then answer the questions that follow.

1. How many unit cubes would fit in a single layer at the bottom of the box?

2. How many identical layers of unit cubes could be stacked in the box?

3. What is the volume and surface area of the box?

Area and Perimeter

The figure below represents a rectangle with a length of 48 centimeters and a height of 16 centimeters. A semicircle with a radius of 8 centimeters has been cut out of each end of the rectangle. Find the area and perimeter of the shaded region of the figure.

Storing DVDs

Malik wants to make an open-top cardboard box to fit on a certain shelf in his closet to store his DVD collection. He has drawn the sketch below of the box that he needs.

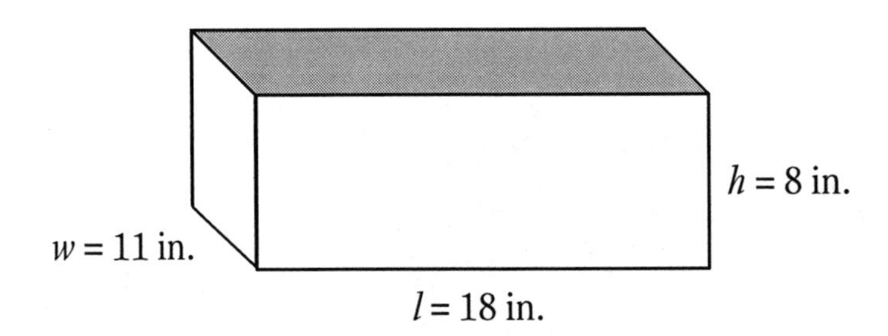

1. How much cardboard will Malik need to create this box?

2. If a typical DVD package measures 7.5 inches × 5.25 inches × 0.5 inches, how many DVDs can Malik expect to store in the box he makes?

3. Malik's dad has brought him a piece of cardboard that measures 2 feet by 3 feet. Is this enough cardboard for the project? Could Malik cut all the pieces that he needs? Explain why or why not.

NAME:

GEOMETRY
CCSS 7.G.6

Storing Trash

According to the U.S. Environmental Protection Agency, U.S. residents, businesses, and institutions made more than 250 million tons of solid municipal waste in 2010. Every ton of waste takes up 2.5 cubic yards of space. Andrew and Todd are wondering how many math classrooms like theirs would be needed to store all that garbage. Their classroom is 45 feet long, 30 feet wide, and 12 feet tall. How many rooms would be needed to store the garbage?

Beach Volleyball

Beach volleyball is played on beaches around the world and is a growing sport in the United States and Canada. The ball used for beach volleyball is a little larger in diameter than a traditional volleyball. It is usually brightly colored compared to the white ball used in indoor games. The Victor Volleyball Company packages its individual beach volleyballs in display cartons that measure 1 foot on each edge. It then ships 12 boxed volleyballs per carton to sporting goods stores all over the world.

1. Find the dimensions of all the different possible shipping cartons that the Victor Volleyball Company could use for exactly 12 balls.

2. Find the surface area of each shipping carton.

3. Which carton needs the least amount of cardboard?

4. Imagine that the Victor Volleyball Company has a greater demand for volleyballs and decides to package 24 boxed volleyballs in a carton. How much packaging material will the company need to create the box with the least material? How much more material is this than the least amount needed for shipping 12 balls?

Pasta Primavera or Chicken Teriyaki?

There are 875 students in Crabtree Junior High. Mrs. Snodgrass, the head of the cafeteria, wanted to add a new entrée to the lunch menu. She was considering chicken teriyaki, beef and broccoli, or pasta primavera. She knew that asking all 875 students would be the most accurate way to find out what to serve. But she also knew that it would be rather difficult to get all 875 students to actually stop and fill out the survey questionnaire.

Mrs. Snodgrass decided that she would ask a group of students to fill out the questionnaire, rather than try to get every student.

How many students should she ask? Would 5 be enough? Would 500 be too many? What do you think? Discuss your answer with a partner. Be prepared to share your thinking with the class.

In the Median

The **median** is the middle value in a set of numbers ordered from smallest to largest. The **mode** is the term that appears most often in a set of numbers. The **maximum** is the largest value in a set of numbers, while the **minimum** is the smallest value. The **range** is the difference between the largest and smallest values of a set of numbers.

1. Here are the math test scores for 12 students.

   70, 90, 90, 40, 80, 60, 50, 20, 40, 30, 20, 40

   What is the range? ________

   What is the minimum? ________

   What is the mode? ________

   What is the maximum? ________

   What is the median? ________

2. Louie kept track of how many hours he worked per week. Here are the hours he worked each week for 12 weeks.

   26, 18, 20, 17, 24, 32, 22, 25, 26, 30, 26, 25

   Find the range of hours. ________

   Find the minimum number of hours. ________

   Find the maximum number of hours. ________

   Find the median number of hours. ________

3. Sarah babysat for seven days in a row. For each day, she earned the following:

Day 1	Day 2	Day 3	Day 4	Day 5	Day 6	Day 7
$15	$20	$25	$30	$25	$25	$35

   What is the range? ________

   What is the mode? ________

   What is the minimum? ________

   What is the maximum? ________

Testing, Testing

In order to make any sense of the data that you collect, you have to organize it first. Below are the scores that 16 students received on their most recent math test:

80, 99, 87, 63, 74, 73, 75, 71, 90, 87, 83, 91, 61, 78, 72, 74

1. Put the data in order from least to greatest.
2. If you divide the data into four equal groups, how many test scores are in each group?
3. What were the scores of the students in the first group (the lowest scores)?
4. What were the scores of the students in the second group?
5. What were the scores of the students in the third group?
6. What were the scores of the students in the fourth group (the highest scores)?

Ring Toss

Shareen and her brother Samir like to go to carnivals and fairs in the summertime. They enjoy the games and the rides. One day, they watched a ring-toss game in which there were old glass soda bottles standing on a wooden platform. The attendant was encouraging Samir to play the game. He said that Samir had a 50% chance of getting any ring to fall over a bottle because the ring will either go on the bottle or fall off. Do you think that Samir and Shareen should believe the attendant? Why or why not? Explain your thinking.

Coin Chances

Solve each probability problem that follows.

1. Alicia tossed a quarter 5 times in a row. It landed tails up 5 times in a row. What is the probability that it will land tails up when she tosses it again? Explain your thinking.

2. Daryl has 5 coins worth exactly 27 cents in his pocket. What is the probability that 1 coin is a quarter? What is the probability that 3 coins are nickels?

3. Make up another probability question about Daryl's coins. Provide the answer, and be ready to explain your thinking.

Pairing Socks

Jamal does not like to take the time to pair his socks after he takes them out of the dryer, so he just throws them into his sock drawer. He knows he has 4 black socks, 2 blue socks, 2 white socks, and 2 tan socks in his sock drawer. One morning he needs to dress in a hurry and wants to wear his tan socks. Without turning on his bedroom light, he reaches in and pulls out one sock.

1. What is the probability that the sock he pulls out is tan?

2. What is the probability that he will get a black or a blue sock on the first try?

3. If he pulls out a tan sock the first time, what is the probability that he will get a tan sock on his next try?

4. What is the probability that Jamal will get a matching pair of socks of any color if he makes two successive selections?

Tossing Coins

Use what you know about probability to answer the questions that follow.

1. If you toss a nickel and a dime together, what is the sample space (or all the possible results that you could get)?

2. What is the probability that you would get heads for the nickel and tails for the dime on the same toss?

3. What is the probability that one coin will be heads and one will be tails?

4. What is the probability that *at least* one coin will be tails?

NAME:

STATISTICS AND PROBABILITY
CCSS 7.SP.8a, 7.SP.8b

Ice Cream Cones

Frosty Freeze features 9 different ice cream flavors each Wednesday. How many different 2-scoop cones could you order if you did not repeat flavors? What if you allowed for 2 scoops of the same flavor as well? Justify your thinking. What if the order of scoops matters? What if order does not matter?

Spinning for Numbers

Look at the spinner below. If it is spun, find each probability given.

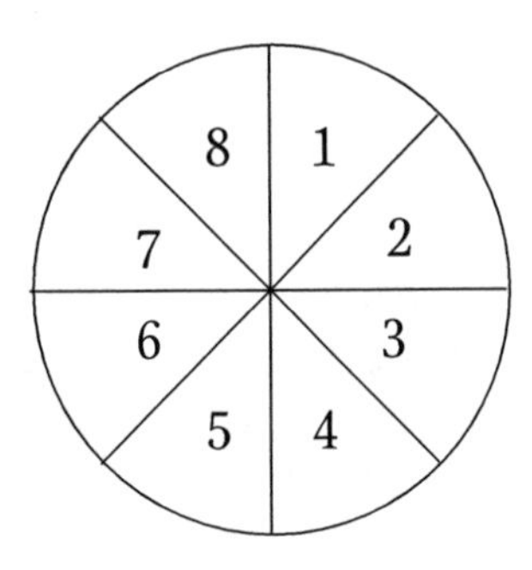

1. *P*(a factor of 12)

2. *P*(a multiple of 3)

3. *P*(9)

4. *P*(a prime number)

5. *P*(an even number)

6. *P*(neither a prime nor a composite number)

Answer Key

Ratios and Proportional Relationships

Balancing a Milk Bottle, p. 1

3.44 miles per hour

Check It Out, p. 2

Answers will vary. The graph is a straight line at an angle. The information on the *x*-axis could be the time the car has traveled. The information on the *y*-axis could be the distance covered. However, the information could be reversed on either axis. Students may say that the graph makes sense because the car is moving at a constant rate.

Comparing Rectangles, p. 3

1. 8 cm^2
2. 72 cm^2
3. 9 times larger (this is found by finding the ratio of the two areas [72/8]). Students may say that they expected the area to be three times larger (since the length and width are three times larger). Encourage them to realize that since *both* the length and the width are three times larger (and they are multiplying these larger numbers), the area will end up being $3 \bullet 3$ or 9 times larger.

Input and Output, p. 4

1. Multiply the input by 2 to get the output.
2. Subtract 4 from the input to get the output.
3. Divide the input by 2 to get the output.
4. Multiply the input by 2 and then add 3 to get the output.

Pizza Pizza Pizza, p. 5

1. The numbers increase by 1.
2. The numbers increase by 1.50.

Tree Height, p. 6

1. The tree's shadow is 56 feet long.
2. The tree is 25 feet tall.

Vanishing Wetlands, p. 7

The original area on the grid is approximately 52 square units. The new area is approximately 34 square units. This represents a 34.6% decrease in area.

Percent Increase or Decrease, p. 8

1. growth; ratio = 3; 300% increase
2. growth; ratio = 1.1; 110% increase
3. decay; ratio = .6; 60% decrease
4. decay; ratio = .2; 20% decrease
5. decay; ratio = .38; 38% decrease

Bread Prices, p. 9

1. 32%
2. $1.37/loaf
3. The 2010 price is 32% greater. If the trend were continuing, one might assume that the percent increase would be closer to 16%.

The Number System

Chip-Board Integers I, p. 10

Chip board 1 should have 6 black chips. Chip board 2 could show the 6 black chips with 6 "zeros" included. Chip board 3 would show 6 white chips remain after 12 black chips are removed. So, $(-6) - (-12) = +6$.

Working with Integers I, p. 11

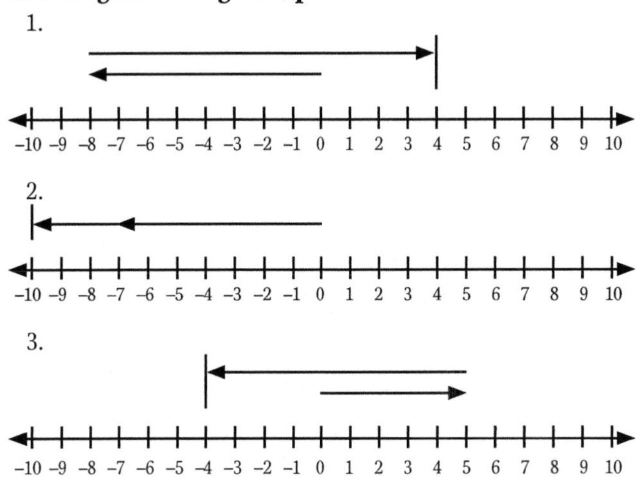

Working with Integers II, p. 12

1. $(-9) + 7 = (-2)$
2. $(-4) + (-6) = (-10)$
3. $7 + (-14) = -7$

Chip-Board Integers II, p. 13

1. $5 + (-8) = (-3)$
2. $(-6) + 10 = 4$
3. $(-8) + 3 = (-5)$

Chip-Board Integers III, p. 14

1. Chip board 1: 5 black chips; Chip board 2: 5 black chips and 9 white chips; Chip board 3: 4 white chips
2. Chip board 1: 12 white chips; Chip board 2: 12 white chips and 7 black chips; Chip board 3: 5 white chips
3. Chip board 1: 8 black chips; Chip board 2: 5 black chips crossed out or removed; Chip board 3 (or 2): 3 black chips

Integer Practice, p. 15

1. −18
2. 33
3. −21
4. −5
5. −1/4
6. 7
7. −2
8. −5
9. −7
10. −2.2
11. $-3\frac{4}{5}$
12. 187/10

Make a True Sentence, p. 16

Answers will vary. Sample answers: $(12 + 34) + (5 \times 6) + 7 + 8 + 9 = 100$; $123 - 4 - 5 - 6 - 7 + 8 - 9 = 100$

10 × 10 Grids, p. 17

1. $\frac{60}{100} = \frac{6}{10} = \frac{3}{5} = 0.6 = 60\%$
2. $\frac{55}{100} = \frac{11}{20} = 0.55 = 55\%$

It's Completely Rational, p. 18

1. $9.09
2. 11/8 = 1 3/8 pizzas or 11 slices
3. $3.32

Four 4s, p. 19

$1 = \frac{4}{4} - (4 - 4)$; $2 = \frac{4}{4} + \frac{4}{4}$; $3 = \frac{(4+4+4)}{4}$; $4 = (4 - 4) \times 4 + 4$; $5 = (4 \times 4 + 4)/4$; $6 = (4 + 4)/4 + 4$; $7 = (4 + 4) - 4/4$; $8 = 4 + 4 + 4 - 4$; $9 = 4 + 4 + 4/4$; $10 = (44 - 4)/4$

Numbering Pages, p. 20

There are 1,024 pages: 9 one-digit pages, 90 two-digit pages, 900 three-digit pages, and 25 four-digit pages.

A Circle of Students, p. 21

1,568 students; $(791 \times 2) - (2 \times 7) = 1{,}568$

Compressing Trash, p. 22

$4.4 \times 0.75 = 3.3 \times 4 \times 365 = 4{,}818$ pounds/50 = 96.36 cubic feet

Expressions and Equations

Perimeter Problem, p. 23

1. $2(5) + 2(4) + 4(2.5) = 28$ meters
2. $P = 2M + 2N + 4L = 2(M + N) + 4L$
3. Answers will vary.

Thinking Around the Box, p. 24

1. $S = 4L + 4W + 4H = 4(L + W + H)$
2. $A = 2LW + 2LH + 2WH = 2(LW + LH + WH)$

Simplifying Expressions, p. 25

1. $8x - 40$ or $4(2x - 10)$
2. $5x^2 - 6x + 13x - 10 = 5x^2 + 7x - 10$
3. $4x + 20 - 6 + 12x = 16x + 14$
4. $30 - 5x + 5x + 5 = 35$
5. $-5x^2 + 3x + 5$ or $5(1 - x^2) + 3x$

Guess and Check, p. 26

1. Peter is 11, and Veronica is 13.
2. José has $7, and Sofia has $28.

Geometry

Scale It Up, p. 27

Students should sketch a square and label each side 20 cm.

Hiking a National Park, p. 28

1. 1.12 km
2. 0.87 km
3. 1.52 km

Can You Do It?, p. 29

Students will mostly likely not be able to figure out how to duplicate an angle. That's okay. This is just a preview to a unit on geometric constructions with a compass and a straight edge.

Angles and Triangles, p. 30

1. 25 degrees. It is half of $\angle EBF$, which is 50 degrees.
2. 25 degrees. It is half of $\angle EBF$, which is 50 degrees.
3. Answers will vary. Sample answer:
   Yes, I think they are the same length because points E and F are the same distance from B, and point G is the same for both segments.
4. Yes, same sides, same angles

Slice It Up, p. 31

1. circle (largest one)
2. circle (smallest one)
3. circle (middle-sized one)
4. triangle

Considering Pizza Costs, p. 32

1. One 12-inch pizza equals four 6-inch pizzas.
2. One 18-inch pizza equals nine 6-inch pizzas.

Pondering Pizza Prices, p. 33

An area pricing model would be best. Explanations will vary.

Circles and Squares, p. 34

1. $A = 100 - (100 \times 3.14/4) = 21.5$ square units
2. $[(100 \times 3.14) - 49 \times 3.14]/4 = 40$ square units

Perimeter Expressions, p. 35

1. no
2. yes
3. no
4. yes
5. no
   Circumference of circle, c, is 2, but Rajah is given a semicircle. Hence, the circumference of the semicirle is πr. The rectangle has 3 sides to consider for the perimeter. The length is $2r$ and the height is r. Thus, the perimeter of the rectangle is $2h + l = 2r + 2r = 4r$. The total perimeter of the figure is $4r + \pi r$, or $r(4 + \pi)$.

Cynthia's Cylinder, p. 36

area of rectangle = 75.36 square inches; total surface area = 89.52 square inches

Complementary and Supplementary Angles, p. 37

1. 53°
2. 39°
3. 45°
4. 64°
5. 146°
6. 60° and 120°

Volume and Surface Area I, p. 38

surface area = 198 square in.; volume = 162 cubic in.

Volume and Surface Area II, p. 39

surface area = 261 square feet; volume = 390 cubic feet

A possible net for the figure:

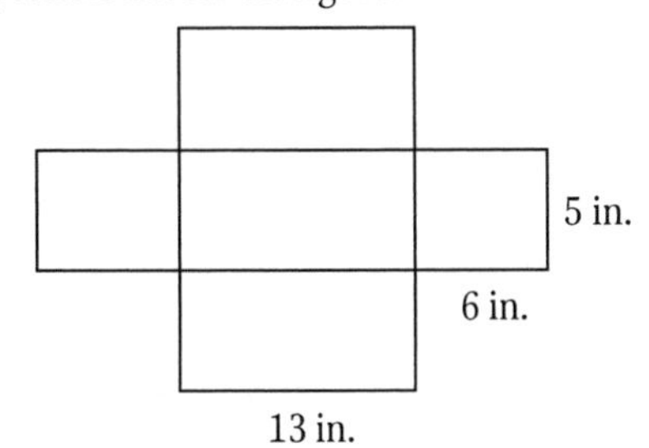

A Rectangular Box, p. 40

1. 15 unit cubes
2. 7 layers
3. 105 in^3; 142 in^2 (open box)

Area and Perimeter, p. 41

perimeter = 146.24 centimeters; area = 567.04 square centimeters

Storing DVDs, p. 42

1. 662 square inches
2. 72 DVDs
3. Yes; answers will vary.

Storing Trash, p. 43

classroom = 600 cubic yards; 250 million tons × 2.5 cubic yards divided by 600 cubic yards ≈ 1,042,000 classrooms

Beach Volleyball, p. 44

1. $1 \times 1 \times 12$; $1 \times 2 \times 6$; $1 \times 3 \times 4$; $2 \times 2 \times 3$
2. 50 square feet; 40 square feet; 38 square feet; 32 square feet
3. the carton that is $2 \times 2 \times 3$
4. The carton that is $2 \times 3 \times 4$ needs 52 square feet of cardboard, 20 square feet more than a 12-ball carton.

Statistics and Probability

Pasta Primavera or Chicken Teriyaki?, p. 45

Answers will vary. Students should recognize that 5 students would not be enough and 500 would probably be too many. There is no "right" answer to this question. However, get students talking about what would be a reasonable number of students to survey (perhaps 100).

In the Median, p. 46

1. range = 70, mode = 40, median = 45, minimum = 20, maximum = 90
2. range = 15, minimum = 17, maximum = 32, median = 25
3. range = $20, minimum = $15, mode = $25, maximum = $35

Testing, Testing, p. 47

1. 61, 63, 71, 72, 73, 74, 74, 75, 78, 80, 83, 87, 87, 90, 91, 99
2. 4 test scores
3. 61, 63, 71, 72
4. 73, 74, 74, 75
5. 78, 80, 83, 87
6. 87, 90, 91, 99

Ring Toss, p. 48

It is not likely that this is a 50% probability. Students might draw diagrams that show by geometric probability that the areas of winning and losing are not half and half.

Coin Chances, p. 49

1. $P(T) = 1/2$
2. P(quarter) = 0; P(nickels) = 0
3. Answers will vary.

Pairing Socks, p. 50

1. P(tan) = 2/10
2. P(black or blue) = 6/10
3. P(tan second try) = 1/9
4. P(any pair) = 18/90

Tossing Coins, p. 51

1. N(H), N(T), D(H), D(T)
2. 25%
3. 50%
4. 75%

Ice Cream Cones, p. 52

There are 72 different cones possible without repeated flavors, or 81 different cones possible if flavors are repeated. There are 36 different cones possible if order doesn't matter for different scoops, and 45 different cones if doubles are allowed.

Spinning for Numbers, p. 53

1. 5/8
2. 2/8
3. 0
4. 4/8
5. 4/8
6. 1/8